Victoria & Albert Museum

Limoges Painted Enamels

Roger Pinkham

London Her Majesty's Stationery Office 1974

The painting of enamels in preference to the *champlevé* process probably began about 1450. Enamelling was a workshop and family concern and one finds generations of Pénicauds, Limosins, de Courts, Laudins and Nouailhers active over three centuries. Using wood-cuts and metal engravings for subjects, artists now painted copper plaques in layers, fixing the colours in a succession of firings. Apart from the use of opaque colour enamel was made more translucent with the aid of metal foil, and gilding, and 'jewels' on the surface enriched the effect. The Museum collection is particularly good in these early colourful enamels, amongst which those by Jean I Pénicaud are extremely skilful (pl. 2). The scenes were invariably religious (pl. 1) with the Life of Christ as the predominant theme (pl. 3). The first period ends with the dissemination of Renaissance iconography by the School of Fontainebleau, enamellers turning now from Northern engravers like Dürer to Italian models like Raphael and Primaticcio, by way of such engravers as Raimondi, Caraglio and subsequent copyists. Polychrome painting was gradually replaced by painting in tones of black and white (*grisaille*) (pl. 5) in the general trend to greater subtlety of technique (cover plate and pl. 3) about 1530. By this time the standing of the craft was high : Léonard Limosin became the first of several enamellers to serve the French monarchy, his court portraits (pl. 4) and *grisaille* work showing the greatest competence. Whereas in the earliest phase religious plaques comprised the usual type of enamel, the decorative idiom of the mid century encouraged a diversity of forms, such as standing dishes, caskets, salts, plates and candlesticks. *Grisaille* enamellers like Pierre Reymond began to increase realism by the addition of flesh tones (pl. 6), an effect that remained when fashion demanded polychrome again with the use of Mannerist subjects about 1550. By this date Italian models were not needed, since a generation of native artists could supply designs, some of the most brilliant utilized at Limoges being by Delaune, Mignon, Boyvin and Du Cerceau. The number of techniques now used surpassed those of the earlier periods, so that *grisaille*, coloured, opaque and translucent enamels may be found together on one piece, here illustrated by the work of Suzanne de Court (pl. 7). That such articles were made solely for decoration is obvious and is in line with the tradition at Limoges, and that they were often commissioned by the rich is evident from the large number of extant pieces bearing heraldic devices (pl. 8).

At the end of the sixteenth century the fortunes of Limoges ebbed as Paris increasingly became the magnet for apprentices and commissions, though much work continued at Limoges, with the Laudin and Nouailher families continuing manufacture into the next century. However, the range of enamels now was very much reduced, consisting mainly of religious plaques that instance the revival of the faith and the popularity of certain saints like St Francis Xavier. The Laudin (pl. 9) and Nouailher workshops maintained painterly standards, but much else became trite.

During the last hundred years the craft has never equalled its distinguished past despite the odd name—Dalpayrat, Popelin, for instance.

1 Triptych. Centre, the Annunciation; left, Louis XII of France and St Louis with the arms of France; right, Anne of Brittany and St Anne with the arms of France and Brittany. Painted and gilt enamels on copper. By the Master of the Louis XII Triptych. About 1500. H. 17⅜ in.; w. 19⅛ in. 552–1877

2 Plaque. The Crowning with Thorns, after a print by Dürer. Painted and gilt enamels on copper. By Jean I Pénicaud; signed. About 1520. H. 9¼ in.; w. 7¼ in. 637–1870

3 Plaque. The Flight into Egypt, after a print by Dürer. Painted enamels and gilding on copper. By Jean II Pénicaud. About 1540. H. $4\frac{7}{8}$ in.; w. $5\frac{5}{8}$ in. C.199–1937

4 Portrait, with figures and cameos in *grisaille*. Charles de Guise, Cardinal of Lorraine. Painted enamels on copper. By Léonard Limosin. Mid 16th century. H. 29 in.; w. 22⅝ in. 551–1877

5 Triptych. St John the Baptist Preaching, after a print by Jean Mignon. Painted *grisaille* enamels and gilding on copper. By M. D. Pape. About 1570. H. 12½ in.; w. 20½ in. 4402, 4403–1857, 544–1893

6 Standing dish. The Evils of Drunkenness. Painted enamels and gilding on copper. By Pierre Reymond ; signed. About 1540. H. $6\frac{1}{8}$ in. ; w. $7\frac{5}{8}$ in. 478–1873

7 Plate. Abraham's Servant meeting Rebekah at the Well, after a print by Bernard Salomon. Painted *grisaille*, and translucent enamels, with gilding on copper. By Suzanne de Court; signed. About 1590. W. 7⅝ in. C.491–1921

8 Casket. Top, a stag hunt. Front and back of lid, Triumph of Bacchus and Ceres, in a chariot drawn by lions. Ends of lid, female figures representing Asia and Africa, from prints by Delaune. On the back, the monogram AA intertwined, the device of Anne of Austria, who married Louis XIII in 1615. Painted enamels with gold and silver decoration on copper. Signed I. L.; probably for Jean I Limosin. About 1620. H. 8$\frac{7}{8}$ in.; l. 12$\frac{3}{4}$ in.; w. 6 in. 13–1864

9 Plaque. Pan and Syrinx, after a print by H. Goltzius. Painted *grisaille* enamel and gilding on copper. By Jacques I Laudin ; signed. Mid 17th century. H. $6\frac{1}{2}$ in. ; w. 8 in. 2049–1855